YOUR KNOWLEDGE HAS VALUE

- We will publish your bachelor's and master's thesis, essays and papers

- Your own eBook and book - sold worldwide in all relevant shops

- Earn money with each sale

Upload your text at www.GRIN.com
and publish for free

Marina Cuvilceva

Aus der Reihe: e-fellows.net stipendiaten-wissen

e-fellows.net (Hrsg.)

Band 468

Explaining the situation of German theatres

Analysis of Performances and Visitors

GRIN Verlag

Bibliografische Information der Deutschen Nationalbibliothek:

Die Deutsche Bibliothek verzeichnet diese Publikation in der Deutschen National-
bibliografie; detaillierte bibliografische Daten sind im Internet über http://dnb.d-
nb.de/ abrufbar.

Dieses Werk sowie alle darin enthaltenen einzelnen Beiträge und Abbildungen
sind urheberrechtlich geschützt. Jede Verwertung, die nicht ausdrücklich vom
Urheberrechtsschutz zugelassen ist, bedarf der vorherigen Zustimmung des Verla-
ges. Das gilt insbesondere für Vervielfältigungen, Bearbeitungen, Übersetzungen,
Mikroverfilmungen, Auswertungen durch Datenbanken und für die Einspeicherung
und Verarbeitung in elektronische Systeme. Alle Rechte, auch die des auszugsweisen
Nachdrucks, der fotomechanischen Wiedergabe (einschließlich Mikrokopie) sowie
der Auswertung durch Datenbanken oder ähnliche Einrichtungen, vorbehalten.

Imprint:

Copyright © 2011 GRIN Verlag GmbH
Druck und Bindung: Books on Demand GmbH, Norderstedt Germany
ISBN: 978-3-656-23501-9

This book at GRIN:

http://www.grin.com/en/e-book/197208/explaining-the-situation-of-german-theatres

GRIN - Your knowledge has value

Der GRIN Verlag publiziert seit 1998 wissenschaftliche Arbeiten von Studenten, Hochschullehrern und anderen Akademikern als eBook und gedrucktes Buch. Die Verlagswebsite www.grin.com ist die ideale Plattform zur Veröffentlichung von Hausarbeiten, Abschlussarbeiten, wissenschaftlichen Aufsätzen, Dissertationen und Fachbüchern.

Visit us on the internet:

http://www.grin.com/

http://www.facebook.com/grincom

http://www.twitter.com/grin_com

Martin-Luther-University Halle-Wittenberg
School of Economics and Business
Chair for Econometrics

Bachelor-Seminar SS 2011: Explaining the situation of German theatres

Topic I: Analysis of Performances and Visitors

Analysis of German theatres in total and in groups of large, middle and small cities

15.07.2011

Marina Cuvilceva
Business Economics

Table of Contents

List of Figures III
List of Tables IV
1   Introduction 1
2   Capacity 1
    2.1   Minimum and maximum capacity in each group 2
    2.2   Mean comparison 3
    2.3   Kruskal-Wallis test of capacity 4
    2.4   Mean capacity in each group 5
3   Performances 6
4   Visitors 8
    4.1   Regarding the type of a ticket 8
    4.2   Regarding the type of a performance 10
    4.3   Mean comparison for visitors 11
        4.3.1   Kruskal-Wallis test of opera visitors 12
        4.3.2   Kruskal-Wallis test of puppet theatres 13
5   Methodology 14
6   Conclusion 15
References 17

List of Figures

Figure 1: Minimum and maximum capacity in each group..2
Figure 2: Kruskal-Wallis test results for capacity 5
Figure 3: Mean capacity in each group 6
Figure 4: Numbers for each type of performance..6
Figure 5: Comparing the number of each type of performance between the groups 7
Figure 6: Numbers for each type of a ticket..9
Figure 7: Comparing the number of each type of a ticket sold between the groups 9
Figure 8: Numbers for visitors with regard to the type of performance...................................10
Figure 9: Comparing the number of visitors with regard to the type of performance between the groups 11
Figure 10: Kruskal-Wallis test results for opera visitors...13
Figure 11: Kruskal-Wallis test results for visitors of puppet theatres 14

List of Tables

Table 1: ANOVA results for capacity 4
Table 2: Summary statistic for opera visitors 12
Table 3: Summary statistic for visitors of puppet theatres 12

1 Introduction

This essay will review the information surveyed on current data from Theaterstatistik 2008/2009 collected by Deutscher Bühnenverein. The data that was given there consists of 112 German cities, number of inhabitants in these cities, capacity of theatres, i.e. seats in all theatres that are available to visitors per 1000 inhabitants in each city. Moreover, there are 146 theatres given with the number of different types of performances produced in that particular theatre as well as the number of visitors presented in two ways: sold tickets to each type of the performance and regarding the type of a ticket.[1]

The aim of this essay is to summarize data about German theatres as well as to investigate interesting similarities and differences, do mean comparison and discuss the methodology used here.

This paper has been divided into four parts. The first part deals with number of inhabitants of given German cities and capacity of theatres in these cities. The second part is concerned with types of performances presented in these theatres. Third part is about visitors in German theatres, which types of performance are mostly visited and which type of a ticket is usually bought. And 4th part contains discussion about methods used when working with the given data and the way the results of the research were presented.

2 Capacity

After exploring data about inhabitants of the given German cities from the first table "Theaterunternehmen", one could present the summary statistics about inhabitants of each city and capacity of its theatres. So, from 112 given German cities the maximum number of inhabitants is 3 431 675 (Berlin), minimum 11 455 (Dinkelsbühl) and the mean number is 531 533.1. For easier comparison of further data, the cities, their theatres and their numbers should be divided with regard to their size into three groups – large, middle size and small cities. Throughout this essay the term "large city" will be now used to refer to the group of cities that have more than 500 000 inhabitants, for example Berlin, Stuttgart, Bremen. The term "middle city" will be used for a group of cities that have between 500 000 and 100 000 inhabitants, e.g. Bonn, Halle, Ingolstadt. And last but not least, term "small city" will be used to refer to the group of cities with less than 100 000 inhabitants, for instance Dessau, Weimar, Eisleben.

---

[1] Theaterstatistik (2010; p. 5f)

2.1 Minimum and maximum capacity in each group

Taking into consideration previously described division of the cities into three groups, the capacity data could be now summarized for each group. A part of summary statistics could be seen in Figures 1 and 3.

Figure 1: Minimum and maximum capacity in each group

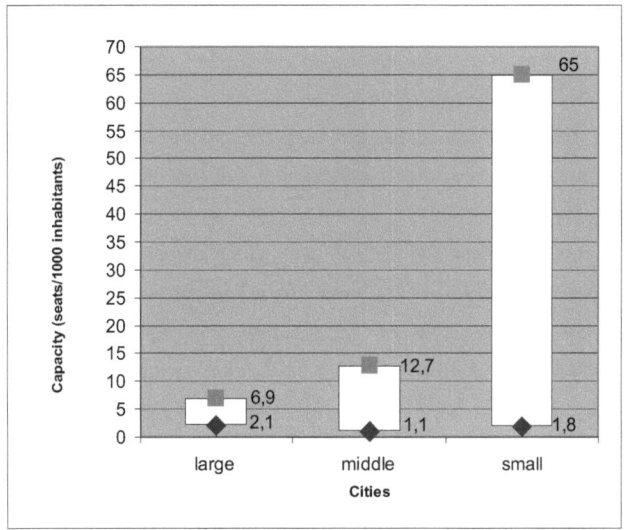

Source: own illustration

First of all, it must be mentioned that the minimum capacity in the given German cities is 1.1 seats per 1000 inhabitants, which is also the minimum capacity of the group of the middle size cities. In other words, in Mühlheim an der Ruhr there is approximately 1 seat available to 1000 inhabitants.

Second, the maximum capacity is 65, which is at the same time the maximum in the group of small cities. Furthermore, 65 is comparatively large number, because the mean capacity of all given cities is only 8.7839 seats. Namely, there are 65 seats available to 1000 people in Anklam, where the number of inhabitants is only 13 423. On the other hand, in the group of large cities almost 7 seats are available to 1000 inhabitants in Nürnberg, where the total number of inhabitants is 503 638.

Moreover, when comparing maximum capacities between the groups, it is visible that the maximum capacity in the group of large cities is much smaller than the maximum capacity of middle size or small cities.

What one could see more from Figure 1 is that the dispersion of the capacity in all three groups is different. So the interval between the minimum and maximum capacity in the group of large cities is comparatively smaller than the interval of middle cities. And distance between minimum and maximum capacity of small cities is very large when comparing to all other groups. This could mean that the capacity of theatres in small cities is not balanced, because there are some cities that have very big number of seats per 1000 inhabitants, for example, Anklam with 65, and some cities which have the minimum capacity number in this group (Aalen 1.8).

2.2 Mean comparison

In addition, one could be interested if the means of these three groups are equal. This is a reason to do ANOVA (analysis of variance), because it is a statistical method for determining the existence of differences among several population means.[2] There are three groups given – 15 large cities, 54 middle size and 43 small cities. Each of them has the capacity of all theatres located in this city. In this case, we have a one-way ANOVA. That is, there is one factor we are looking at across these 3 groups.

Step 1: State the null hypothesis and the alternative hypothesis. The null hypothesis in ANOVA is that the means of the groups are equal.[3] In other words, mean of group of large cities is equal to the mean of group of middle and the mean of the group of small cities. And alternative hypothesis is that means of these three groups are not equal.[3]

In other words, if the hypothesis is true, then the "between group variance" (MSTR) will be equal or close to the "within group variance" (MSE). MSTR (Mean Square Treatments) is the estimate of the population variance based on the differences among the sample means. And MSE (Mean Square Error) is the estimate of the population variance based on the variation within the sample.[3] If, on the other hand, the null hypothesis is not true and differences do exist among the population means, then MSTR will tend to be larger than MSE.[4]

Under the assumption of ANOVA, the ratio MSTR/MSE possesses an F distribution with r-1 degrees of freedom for the numerator and n-r degrees of freedom for the denominator when the null hypothesis is true.[4]

Step 2: Select the level of significance. In this case, 0.05 is chosen.

Step 3: Calculate the F statistic using Excel's Data Analysis. Results are presented in Table 1.

---

[2] Aczel (2006; p. 370)
[3] Lind (2005; p. 395f)
[4] Aczel (2006; p. 383)

Table 1: ANOVA results for capacity

Anova: Single Factor						
SUMMARY						
Groups	Count	Sum	Average	Variance		
Large	15	60,4	4,0267	2,5664		
Middle	54	306,1	5,6685	5,6501		
Small	43	617,3	14,3558	132,2978		
ANOVA						
Source of Variation	SS	df	MS	F	P-value	F crit
Between Groups	2198,5592	2	1099,2796	20,3367	0,0000	3,0796
Within Groups	5891,8919	109	54,0541			
Total	8090,4511	111				

Source: own illustration

Step 4: Interpret the results. As one could see, the mean level of capacity in small cities (14.3558) is higher than that of either middle (5.6685) or large cities (4.0267). And "between group variance" is not equal to "within group variance". According to the test result F equals 20.3367. With a significance level of 0.05, the critical F is 3.0796. Therefore, since the F statistic is greater than the critical value, and the p-value is much smaller than 0.05, the null hypothesis is rejected. From above, the null hypothesis was that all 3 of these groups' means were equal. So, it should be rejected that large, middle and small cities have the same level of capacity.

However, an important problem to mention is that variances of the groups are not equal, so ANOVA should not be used and instead one could use a nonparametric technique called Kruskal-Wallis test.[5]

2.3 Kruskal-Wallis test of capacity

Taking into consideration the last conclusion, one could use the Kruskal-Wallis which is also known as nonparametric one way ANOVA. The Kruskal-Wallis test is an analysis of variance that uses ranks of the observations rather than the data themselves. Although the hypothesis is stated in terms of the distributions of the populations, the test is most sensitive to differences in the locations of the populations. Therefore, the procedure is actually used to test the ANOVA hypothesis of equality of k population means. The only assumptions required for Kruskal-Wallis test are that the k samples are random and independently drawn. [6]

Step 1: Null hypothesis is that all 3 populations have the same distribution. And alternative hypothesis is that not all of them have the same distribution. [6]

---

[5] Aczel (2006; p. 372)
[6] Aczel (2006; p. 665f)

Step 2: 0.05 level of significance could be chosen.

Step 3: Calculate Kruskal-Wallis test statistic with the help of Excel Analyse-it. And the results are presented in Figure 2.

Figure 2: Kruskal-Wallis test results for capacity

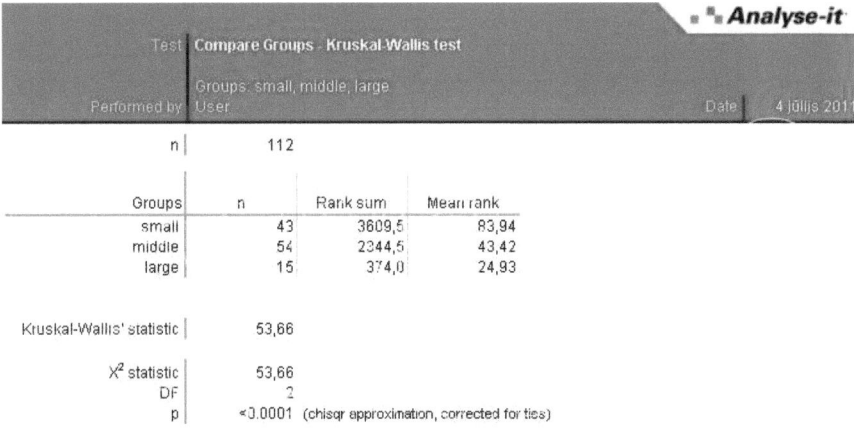

Source: own illustration

Step 4: As one could see the Kruskal-Wallis test statistic is 53.66. And the critical value that could be obtained from chi-square distribution table with k − 1 = 2 degrees of freedom and 0.05 significance level is 5.99. [7] That is why it should be concluded that the null hypothesis is rejected, so all three groups do not have the same level of capacity.

2.4 Mean capacity in each group

In addition, the mean comparison in dot plot in Figure 3 shows that capacity of theatres in small cities is much bigger than in middle or large cities. Hence, capacity of middle size cities is also bigger than the one in large cities. For example, capacity of theatres in Halle (middle size city) is 11.1, which is much larger than 2.7, the capacity of Berlin theatres. Moreover, the capacity of Eisleben from the group of small cities is 21, which is higher than the capacity of any city from the group of large or middle size.

---

[7] Aczel (2006; p. 779)

Figure 3: Mean capacity in each group

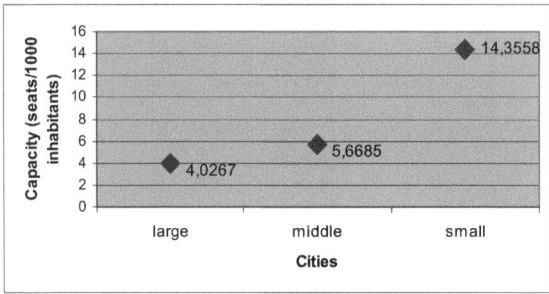

Source: own illustration

3 Performances

The data in the second table of theatre statistics, called "Veranstaltungen", presents the theatres in each city, number of each type of performance produced in it and whether it is produced there at all. Performances are divided into different types, such as opera, dance, operetta, musical, play, children and youth theatres, concert, puppet theatre, guest plays of foreign groups and other performances. The total number of all performances produced in given theatres is 65 508.

As one could see in Figure 4, play is the most often created type of performance in this given German theatres statistics, with 23 732 performances. The second best is children and youth theatre (12 287) and opera is the next with 6 473 performances. In addition, operetta has the least number of performances produced for that type (1 232).

Figure 4: Numbers for each type of performance

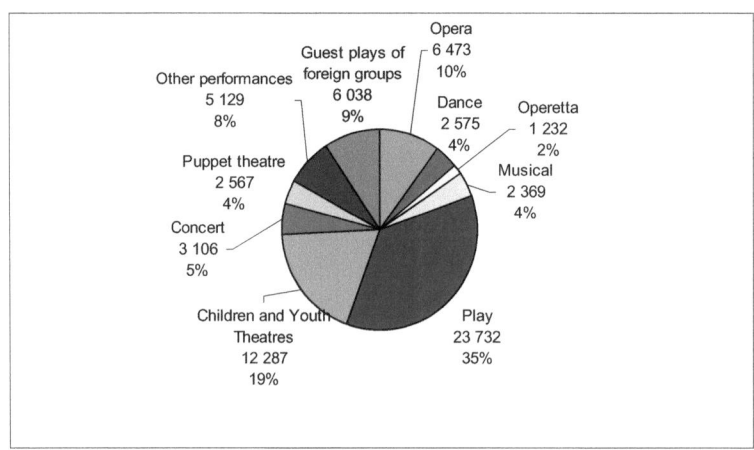

Source: own illustration

Figure 4 shows also the share of each type of performance produced out of the total number. The most often produced type, play, has 35% of all performances produced in given theatres. Children and youth theatres have 19%, opera and guest plays of foreign groups have around 9-10%. Other types have less than 10% out of all performances presented here. Operetta has the smallest share, which is around 2%.

Taking into consideration previously mentioned division of the cities into three groups, the data about number of performances could be also sorted out for these groups. First of all, number of theatres in each of them is different, which is in big cities - 39, in middle size – 63 and in small cities – 44 theatres. As a result, the total number of all types of performances in middle size cities is 31 200, in large – 19 935 and in small cities – 14 373 performances.

Figure 5: Comparing the number of each type of performance between the groups

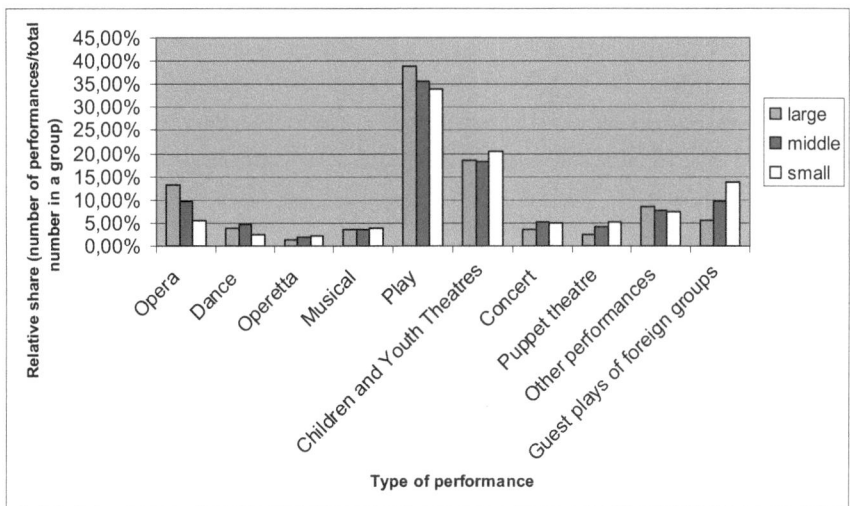

Source: own illustration

Summing up all the numbers of each type of performance for each group and dividing the number for each type of performance by the total number of performances in each group, one could get the results that are presented in Figure 5.

First of all, when comparing the top three performances in each group, one could see that it is almost the same for all cities. First is play, then children and youth theatres, and then opera. However, in small cities the third place goes to guest plays of foreign groups. The least number of performances in all cities belongs to operetta (2%).

Second, play has the biggest relative share in all groups and has significant difference when comparing to the other types of performances. It has almost twice higher number than children and youth theatres, and has even bigger difference when comparing to other types.

Third, the relative share declines almost proportionally when the number of inhabitants falls when exploring the results for opera and play.

Moreover, children and youth theatres have slightly higher share (20%) in the group of small cities when comparing to other groups (18%).

In addition, quite logical result could be seen in the Figure 5, that the highest number of guest plays (14%) belongs to the group of small cities and this number declines when the size of the city rises.

4 Visitors

When exploring the data in the third table of theatre statistics, table "Besucher", one could discover that there were 21 354 646 tickets sold to all performances in given cities. In this table there are numbers for these types of tickets available: full price, season ticket/seat reservation, visitor organizations, children and students tickets, discounted tickets, tickets for employees of theatres and free tickets. There are also given numbers of tickets sold to different types of performance.

4.1 Regarding the type of a ticket

The most popular type of a ticket is, of course, full price ticket (7 841 908), which is 40% out of all tickets sold. Second are season tickets/seat reservations with 19% and the next are children and student tickets with 16%. The least share has tickets given out to employees of the theatres, only 3%, mainly because this part of the visitors is comparatively small.

To compare the number of visitors with respect to the type of the ticket in these three groups mentioned before, one should look at the Figure 7 that presents the relative share of each type of a ticket out of all the tickets sold in that group. The total number of tickets sold in the group of large and middle size cities is almost equal (around 8.7 million). However, the total number of tickets sold in the group of small cities is significantly smaller; it is only 3.9 million tickets.

Figure 6: Numbers for each type of a ticket

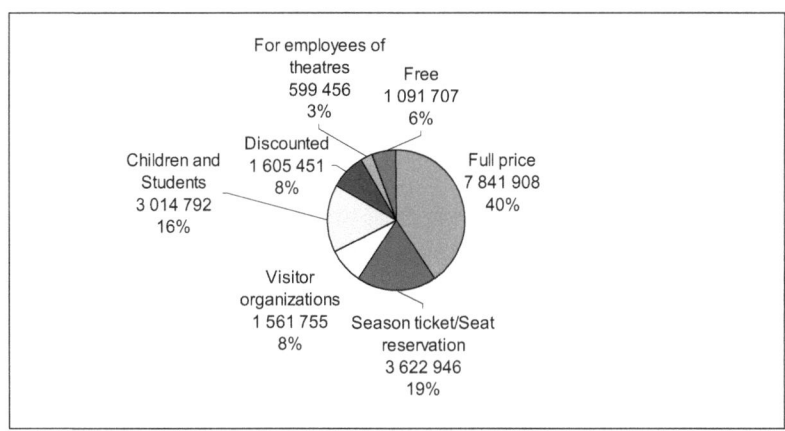

Source: own illustration

First of all, when looking at the Figure 7 one could see that the most often sold type of a ticket is the full price ticket in all three groups. However, large cities are the leaders in this type (40%). Moreover, this type of a ticket has significantly higher share than all other types in all three groups.

Second, there is comparatively bigger number of season tickets sold in middle cities (21%) than in large (15%) or small cities (12%).

Figure 7: Comparing the number of each type of a ticket sold between the groups

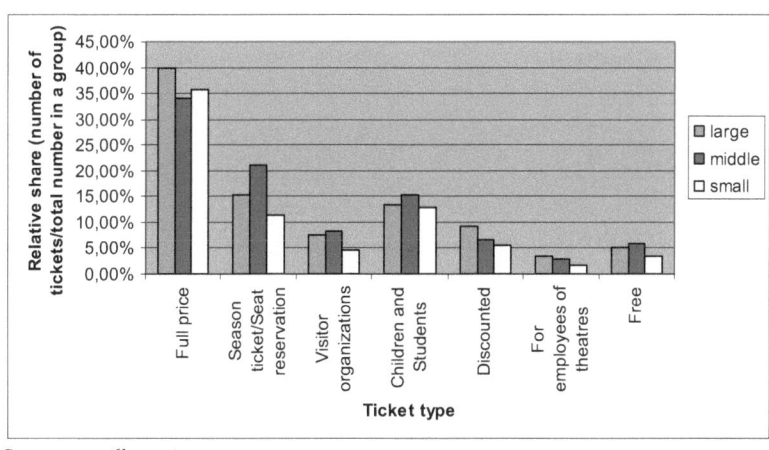

Source: own illustration

Third, there is smaller number of tickets sold to visitor organizations (5%) and free tickets (3%) in small cities when comparing to large or middle cities.

Furthermore, the group of large cities has the highest number of discounted tickets (9%) sold when comparing to other groups. And this number declines when the size of the city declines.

4.2 Regarding the type of a performance

The summary statistics of visitors with respect to the type of performance is shown in Figure 8. First, the top three most popular performances visited in all given cities is the same as the top three of the most often produced performances. These are – play, with 5.6 million visitors, which is 27% out of all visitors, opera (4.4 million, 21%), and children and youth theatres (2.7 million, 12%). All other performances have less than 10% of visitors out of the whole number. The least popular is puppet theatre with 225 540 visitors, which is only 1% of visitors.

Figure 8: Numbers for visitors with regard to the type of performance

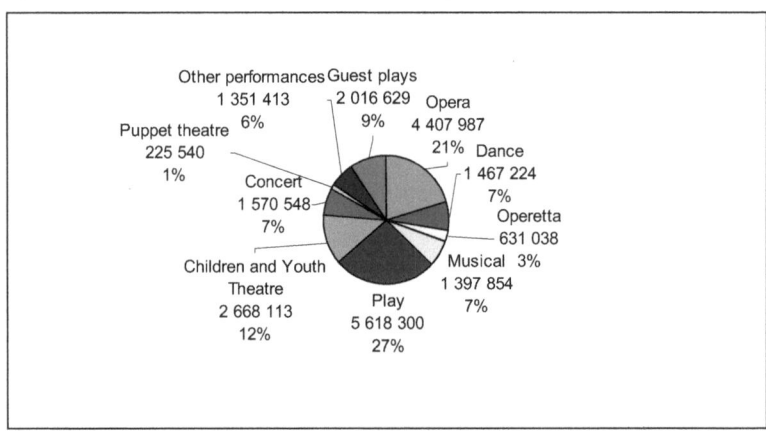

Source: own illustration

When looking at the Figure 9 one could compare the relative share of the visitors of each type of performance in these three groups mentioned before.

First, the top three most visited performances are now different in different groups. In large cities opera has the highest relative share (30%), and then go play (26%) and dance (9%). Inhabitants of middle cities prefer visiting play (28%) to opera (17%) and children and youth theatres (16%). In small cities people prefer visiting guest plays (24%), then play (24%) and children and youth theatres (16%). The least relative share in all three groups belongs to puppet theatres (1%).

Second, there are significant differences in relative shares of visitors of operas between the groups. And the number declines when the size of the city declines. The relative share of the visitors of dance performances also declines when the size of the city declines.

Third, there is a significantly higher relative share of visitors of guest plays in small cities (24%) than in other groups (6%).

Furthermore, the relative share of visitors of children and youth theatres in large cities (8%) is much smaller than in middle or small cities (16%). And the relative share of visitors of concerts and operettas is also smaller in large cities when comparing to other groups.

Figure 9: Comparing the number of visitors with regard to the type of performance between the groups

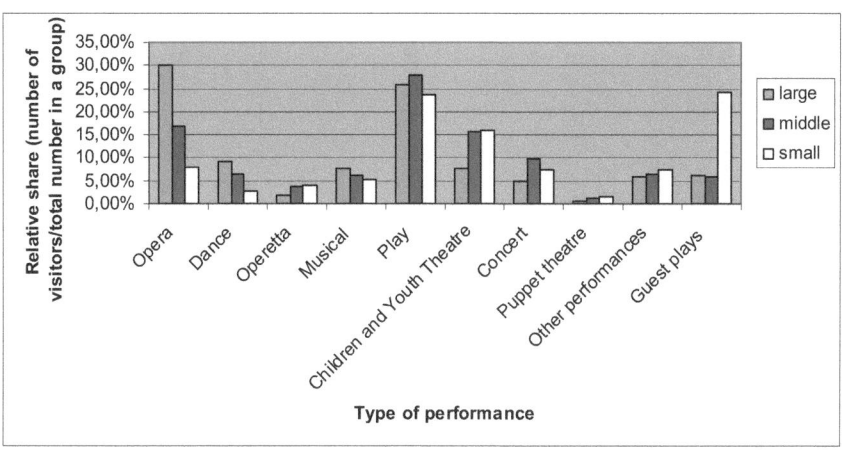

Source: own illustration

4.3 Mean comparison for visitors

In addition, some separately taken types of performance could be tested. For example, let us compare summary statistics of visitors of opera and visitors of puppet theatres between three groups of cities. And one could see from Tables 2 and 3 that the variances again are not equal so the assumption for ANOVA is not met again. That is why one could use Kruskal-Wallis test instead.

Table 2: Summary statistic for opera visitors

SUMMARY				
Groups	Count	Sum	Average	Variance
large	39	2652	68,0000	6193,0000
middle	63	3021	47,9524	1825,6267
small	44	800	18,1818	673,1755

Source: own illustration

Table 3: Summary statistic for visitors of puppet theatres

SUMMARY				
Groups	Count	Sum	Average	Variance
large	39	57083	1463,666667	33176702,23
middle	63	109997	1745,984127	30883322,21
small	44	58460	1328,636364	11473935,63

Source: own illustration

4.3.1 Kruskal-Wallis test of opera visitors

In this analysis there will be again used the same statistics of three groups of cities with their numbers of theatres: 39 in large, 63 in middle size, 44 theatres in small cities. Each of them has data that represents the number of opera visitors and if there were no tickets sold, because there is no opera performance at all in this theatre, then the number of visitors is 0.

Step 1: The null hypothesis in this analysis is that all 3 populations have the same distribution. If the hypothesis is rejected, it should be concluded that these 3 populations do not have the same distribution.

Step 2: 0.05 level of significance could be chosen.

Step 3: Calculated statistic using Excel's Data Analyse-it is shown in Figure 10.

Step 4: When interpreting the results, one could see that the Kruskal-Wallis test statistic (13.22) is bigger than the obtained critical value with 2 degrees of freedom and 0.05 significance level (5.99). Therefore, large, middle and small cities do not have the same level of opera visitors.

Figure 10: Kruskal-Wallis test results for opera visitors

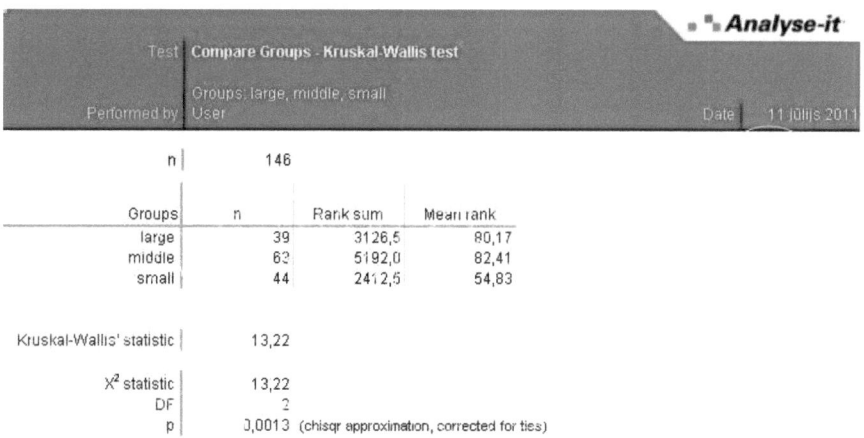

Source: own illustration

4.3.2 Kruskal-Wallis test of puppet theatres

There are the same groups used again, but the numbers that are observed in this section represent the number of visitors of puppet theatres and if there are no puppet theatres in a city then the number is 0.

Step 1: The null hypothesis is that all 3 populations have the same distribution. The alternative hypothesis is that they do not have the same distribution.

Step 2: 0.05 significance level is chosen.

Step 3: The results calculated by Excel's Data Analyse-it are presented in Figure 11.

Step 4: As one could see, the Kruskal-Wallis test statistic (6.14) is bigger than the obtained critical value from the table of chi-square distribution (5.99). And again the null hypothesis is rejected, even though these numbers are almost equal. So the level of visitors of puppet theatres is not the same for these three groups. But if one rounds these results to the whole numbers then the null hypothesis could be rejected. Then one could assume that all three groups have the same level of visitors of puppet theatres.

In this case then when summarizing these two tests conducted before, one could see that the mean level of opera visitors is different across these three groups of cities and mean level of visitors of puppet theatres are equal or almost equal across these groups. In addition, the other proof of that could be Figure 9, where one could see that the differences in numbers of opera visitors are significant between the groups and at the same time the differences in the numbers of visitors of puppet theatres are almost invisible. To conclude, few methods could be used to test and to present the results.

13

Figure 11: Kruskal-Wallis test results for visitors of puppet theatres

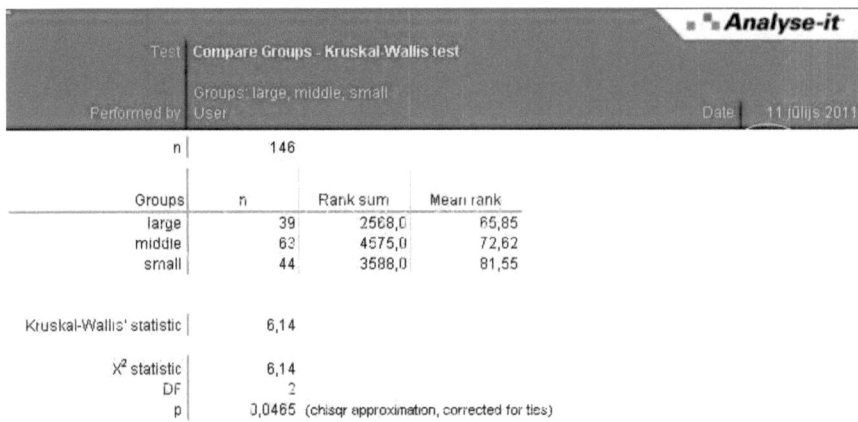

Source: own illustration

5 Methodology

It is now necessary to explain the methods used in this paper.

As the main goal of this essay was to describe given statistics and present interesting results, that is why the best way how to do this is to show the summarized data in different types of graphs.

First of all, there were numerous bar charts used to display categorical data where there is no emphasis on the percentage of a total represented by each category.[8] The main purpose of it was to compare the numbers between the three groups: small, middle and large cities.

Second, there were also few pie charts used to display the data that sums up to a given total.[8] According to opinion of Amir D. Aczel in his book "Complete Business Statistics" a pie chart is the most illustrative way of displaying quantities as percentages of a given total.[8] Nevertheless, few opinions about pie charts are contrary to the previous one. For example, Stephen Few recommends avoiding pie charts. He also said that despite the obvious nature of a pie chart, bar graphs provide a much better means to compare the magnitudes of each part. Because any percentages other than 0%, 25%, 50%, 75% or 100% are difficult to discern, but can be accurately presented in a bar graph.[9] In addition, Edward Tufte, professor of statistics and computer science, author of "The Visual Display of Quantitative Information", writes that pie chart should never be used at all. Moreover, he once said that "the only worse design than

---

[8] Aczel (2006; p. 46)
[9] Few (2007; p. 2)

a pie chart is several of them, for then the viewer is asked to compare quantities located in spatial disarray both within and between pies".[10]

Third, there were also few dot plots used in section 2 to display specific points or exact numbers in each group, mainly for these numbers to be easier to compare.

As Amir D. Aczel stated, a picture or a graph is indeed worth a thousand words, but they can sometimes be misleading. Often, this is where "lying with statistics" comes in, when the data is presented graphically on a stretched or compressed scale of numbers or when there is no scale at all. And this is done with the aim of making the data show what people want them to show. This is an important argument a merely descriptive approach to data analysis.[11]

In contrast to this, as Aczel states statistical tests tend to be more objective than people's eyes and are less exposed to deception as long as the assumptions hold. Statistical inference gives tools that allow evaluating objectively what one should see in the data.[11] Moreover, along with regression analysis ANOVA is the most commonly quoted advanced research method in the professional business and economic literature.[12] However, in this paper the assumption of equal variances does not hold true for ANOVA and that is why the nonparametric Kruskal-Wallis test was used, that requires no assumptions about the shape of the population distributions.[13]

Finally, a problem of the research provided in this paper should be also mentioned. In most of the figures presented here one could notice that the middle cities are the leaders, when comparing them to other groups of cities, for example Figure 5: Comparing the number of each type of performance between the groups. This might be because the sizes of the groups are not equal. In other words, number of theatres in the group of middle cities is 63, which is around 30% larger than the number of theatres in small (44) or large cities (39). Generally speaking, the subjectively chosen division into groups by size could cause the problems of the whole research and might show misleading information.

6 Conclusion

All things considered, this essay provided a short research on statistical data about theatres, performances and visitors of these theatres in Germany. As a result, there were many interesting facts discovered during the research.

---

[10] Tufte (1983; p. 178)
[11] Aczel (2006; p. 47)
[12] Aczel (2006; p. 370)
[13] Aczel (2006; p. 665)

First of all, research of the capacity data in section 2 – minimum, maximum and mean comparison of three groups of cities showed that the mean capacity rises when the number of inhabitants falls and that the highest mean capacity is in the group of small cities. So generally speaking, this could mean that people in small cities have more opportunities of visiting theatres. Moreover, capacity of theatres does not directly depend on the size of the city. In other words, if the city is large it does not mean that there are more theatres or seats in these theatres available to its inhabitants than in small city.

Second, looking at the figures presented in this paper, it could be summarized that the most often produced and most often visited type of performance are almost the same in all groups. These are play, opera, guest plays, children and youth theatres. The least popular type of performance is puppet theatre. The most often bought tickets are full price tickets, season tickets and children and student tickets.

Third, people in large cities most often visit operas and then plays. However, people in middle size cities prefer plays to operas, maybe because operas as separate theatres are usually located in bigger cities. And people in small cities prefer visiting guest plays rather than their local plays.

Finally, the research provided in this paper showed only a small part of information that could be presented from the given statistics. Still some further and more detailed research of German theatres could be very helpful in presenting the general information, getting to know the strengths and weaknesses of German theatres and also in decision making, for example, for government institutions that are concerned with culture, for municipalities, managers of theatres and for private groups of artists.

References

1. Aczel, Amir D., Sounderpandian, J. (2006): Complete Business Statistics, 6th ed., McGraw-Hill, Singapore
2. Deutscher Bühnenverein (2010): Theaterstatistik 2008-2009, Bundesverband der Theater und Orchester, 44, 3-117
3. Few S. (2007): Save the Pies for Desserts, Visual Business Intelligence Newsletter, Perceptual Edge
4. Lind, D., Marchal, W., Wathen, S. (2005): Statistical Techniques in Business and Economics, 12th ed., McGraw-Hill (international edition), Singapore
5. Tufte E. (1983): Visual Display of Quantitative Information, Graphics Press, Cheshire